SET THE PLOUGHSHARE DEEP

SET THE PLOUGHSHARE DEEP

a prairie memoir

Timothy Murphy

woodcuts by Charles Beck

Ohio University Press
Athens

Ohio University Press, Athens, Ohio 45701

Printed in Hong Kong by C & C Offset Printing Co., Ltd.

Ohio University Press books are printed on acid-free paper ⊗ ™

08 07 06 05 04 03 02 01 00 5 4 3 2 1

Library of Congress Cataloging-in-Publication Data

Murphy, Timothy, 1951–

Set the ploughshare deep : a prairie memoir / Tim Murphy ; woodcuts by Charles Beck.

p. cm.

ISBN 0-8214-1321-X (alk. paper) — ISBN 0-8214-1322-8 (paper : alk. paper)

1. Murphy, Timothy, 1951—Homes and haunts—North Dakota. 2. Poets, American—20th century—Biography. 3. Farmers—North Dakota—Biography. 4. Farm life—North Dakota—Poetry. 5. Farm life—North Dakota. 6. North Dakota—Poetry. I. Beck, Charles. II. Title.

PS3563.U76195 Z47 2000

811'.54—dc21

[B]

99-089156

for Vincent R. Murphy

Foreword

You are about to read a memoir in verse, prose, and woodcuts. Though conceived separately by three different people, these works tell a single story, as surely as the rings in neighboring tree stumps tell the story of our harsh climate. North Dakota is prone to every extreme of weather. In mild years the prairie blooms; yet flood and drought, heat and freeze are perennial threats to agriculture. Grandson of a sodbuster, Timothy Murphy has relived a family tradition of boom and bust. After schooling on the East Coast and several years of city life in Minneapolis, he returned to his native soil along the Red River of the North. Here he evolved the distinctive and musical style of his lyric poems, which are as terse as gunshots in duck season.

The poetry and the prose of this memoir accreted for fifteen years until they took their present shape. Moved by his son's account, Vincent Murphy contributed his recollections of the Great Depression. Later the woodcut artist Charles Beck agreed to let a selection of his own works appear. Born east of the Red in Fergus Falls, Minnesota, Beck has lived and worked close to the land all his life. As stark as Murphy's poems, Beck's images hover mysteriously between abstraction and photographic accuracy. At one end of the visual continuum, Beck minutely examines plants as forms in rhythmic repetition. At the other, he draws the eye to a horizon depicted with such depth that it would surely alarm anyone prone to agoraphobia. But always there is the sense of forms behind forms, space beyond space.

Memoir has become such a popular genre in recent decades that courses in memoir writing are now taught for undergraduates. How many of them have lived enough to justify such exercises? A memoir worth reading requires the perspective of age, even if it reflects upon the experiences of youth. But no rule can govern what form the insights of age may take.

In *Set the Ploughshare Deep* the intrinsic form is verse. Just as the Beck woodcuts are poetic in their concern with rhythm and line, the Murphy prose is very nearly poetry. So in a sense *Ploughshare* is one extended poem of the land and the people shaped by that land over a century of settlement.

Alan Sullivan
Fargo, North Dakota
November 1998

Acknowledgments

I wish to thank the editors of the following publications in which poems first appeared: *Hellas, The Epigrammatist, Light, The Formalist, Chronicles, Janus, The Dark Horse, Prop, The Edge City Review.*

Many of these poems also appear in *The Deed of Gift* (Story Line Press, 1998) and in three chapbooks, *The Ant Lion* (R. L. Barth, 1996), *Bedrock* (Aralia Press, 1998), and *Tessie's Time* (Fameorshame Press, 1999).

Thanks to my father for permission to quote his memoir.

Thanks also to my collaborator, Alan Sullivan, who gathered the fragments of this story into a narrative and ghost-wrote transitions wherever the pieces didn't fit.

Finally, a thank you to James O'Rourke and the Rourke Gallery in Moorhead, Minnesota, for their help in reproducing the Beck woodcuts.

S U N F L O W E R S

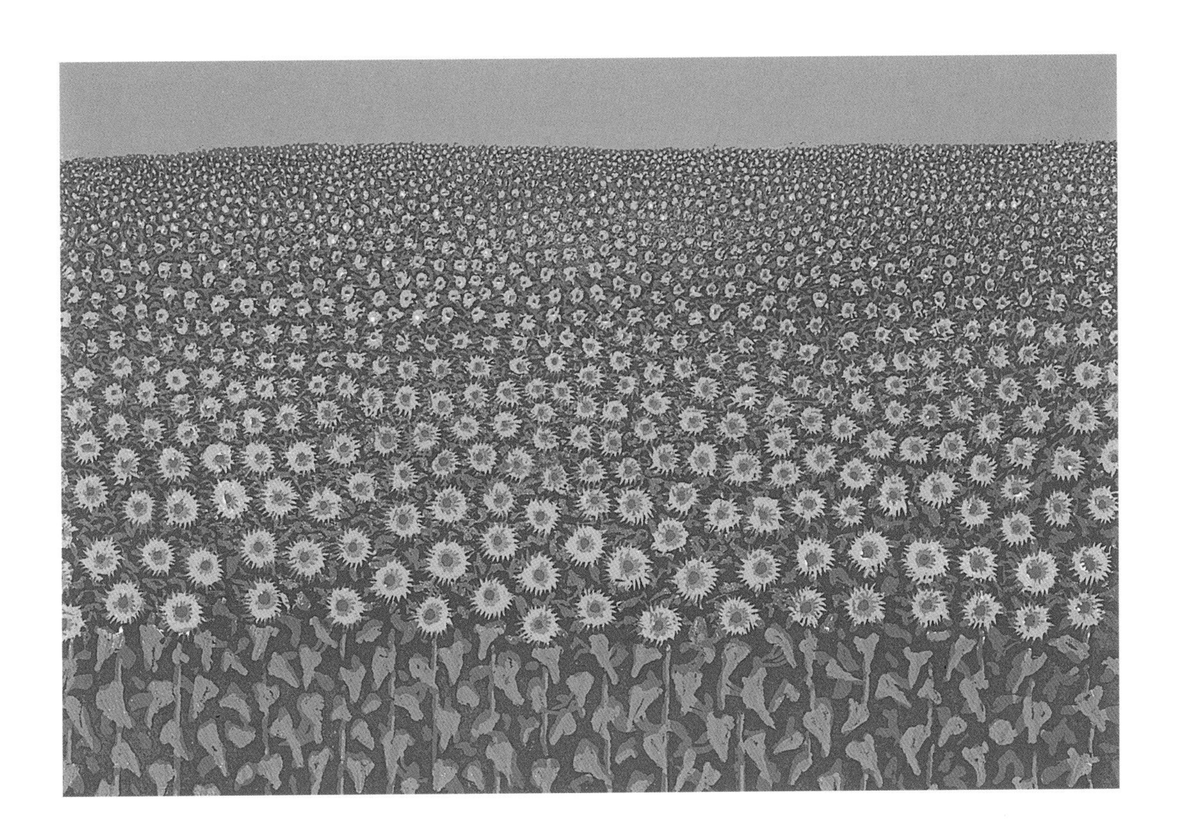

1 9 9 3

The Godless Sky

Airborne, I view the flood.
  Flattened wheat
  blackens in the heat
or smothers under mud.

In the Piper's path I spot
  the farmstead
  on the home spread
my father's father bought.

An uncle stauncher than I
  bears the cross
  of another loss
under the godless sky.

It's hard to imagine a horizon as flat as a ruled line unless you've sailed far out to sea or driven Interstate 29 along the Red River of the North. Planed by an Ice Age lake, our Valley extends hundreds of miles along the border of Minnesota and the Dakotas, widening from thirty to ninety miles as it drops imperceptibly northward toward Winnipeg. To our east rise the maple- and pine-clad hills of northern Minnesota. Buffalo country lies to the west, the rolling rangeland that divides our Canadian watershed from the Missouri. The soil there is arid and alkaline. Eastward the soils vary from peat bogs to glacial till, sandy and stony. But the Valley is black loam. In a good year it's the most productive land in the world. In bad years it's a desert, or a lake.

I was three years out of college when I witnessed the flood of 1975. Our floods usually come in spring, when rains and snowmelt coincide. This one was different. It started on a muggy July afternoon as thunderheads blew up over my grandfather's farm northeast of Moorhead, Minnesota. The storm grew until it covered three counties, and it continued all evening, with hail big enough to kill cattle and tornadoes that tore down barns and silos. The benchland east of the Valley shed the deluge like a roof, and floodwaters guttered onto the farms below, gouging huge gullies in the fields.

I stayed high and dry in town, but the storm nearly wiped me out anyway. As crops rotted in the swamped countryside, my livelihood went with them. In a farm economy even the townspeople take their living from the land. You might think I would want nothing to do with farming after an experience like that. But the 1975 flood was only a bad memory in the drought year of 1976, when there were dust storms and locusts to worry about. Then came 1977, with a bumper crop and hungry Soviets bidding up our prices. I was in clover again. Bust and boom seem all too natural to a homesteader's grandson.

Razing the Woodlot

Here stands the grove our tenant plans to fell.
The homesteaders who planted this tree claim
fled North Dakota when the Dust Bowl came.
Their foursquare farmhouse is a roofless shell;
their tended shelterbelt, a den for fox
and dumpground for machinery and rocks.

The woodlot seeds its pigweed in our loam,
and windstorms topple poplars on the field;
but for a few wasted acres' yield
we'll spare the vixen and her cubs their home
and leave unburied these decaying beams
to teach us the temerity of dreams.

E. J. Murphy was born in upstate New York in 1856, son of a prosperous, second-generation Irish-American farmer. But home was hardscrabble country—rocky, steep, and almost as exhausted as Ireland. At age twenty E. J. went to Wisconsin to seek his fortune in the lumber camps. Later he joined the railroad and arrived with the locomotive at the opening of the Red River Valley. Smitten with this vast swale of loamy, rock-free prairie, he determined to homestead his own land. Back in the camps he accumulated a stake of capital and lumber sufficient for the venture. Heading west again, he purchased a wagon, oxen, seed, and a few rudimentary implements at the thriving rail town of Fargo. He then trekked twenty miles northeast, drove his stakes into a quarter section of prairie, and broke the virgin bluestem with a two-bottom plough. With his imported lumber he built the first frame house in treeless Felton Township.

E. J. had ten children by his first wife, who died (probably of exhaustion) in the fifteenth year of their marriage. When the widower was elected Felton's Superintendent of Schools, he hired a comely schoolmarm thirty-two years his junior, whom he soon married. In the post-war inflationary boom, Grandfather raised a new family on the homestead, which he had leveraged to three thousand acres, emulating the bonanza farmers he'd envied in his youth, and provoking in turn the envy of neighbors with eighty acres and a mule.

The first exuberant wave of settlement had crested long before, but high grain prices sustained large-scale farming despite falling yields and expensive transport. In E. J.'s heyday, disputes between farm cooperatives and the railroads engendered a political movement, the so-called "Non-Partisan League," enshrining a partnership between farmers and the state that now seems increasingly burdensome to all parties. The League disbanded when land values soared to $250 an acre and wheat brought $4 a bushel. Amid such prosperity the conflicts of the past seemed moot until over-production, Black Tuesday, the Smoot-Hawley tariffs, and the ensuing worldwide economic crash dethroned King Wheat. In 1932 the price of a bushel dropped to 14 cents. Felled by a stroke, E. J. left a bundle of debts to his young wife and sons.

Armed only with a parka and a teaching certificate, seventeen-year-old Tessie Buckley had taken her first job in a hamlet near the Canadian border. Bypassed by the railroad, that village no longer exists; but my brother and I have often hunted geese in its vicinity. There Tessie drilled *McGuffey's Reader* into German-speaking youths whose farmer fathers, admiring her lustrous hair, called her "die kleine rote Fuchsen," the little red fox.

Later Tessie took degrees in English and history, unheard-of credentials for a farm girl in 1912. When she assumed a new post in Felton, Tessie saw E. J. sweep into the schoolyard with his fine quarterhorses and pedigreed hounds. How could she refuse an Irish suitor in this valley of Scandinavians?

Tessie had three children before her husband died, broken by the Depression and the specter of insolvency. My father Vince was the eldest. The second child, Vivian, was paralyzed by an incompetent surgeon two years after her father's death, and she was to be Tessie's burden and inseparable companion for nearly half a century. The youngest, Dan, went to war in 1943. Shot down and presumed lost over Germany, Dan survived to farm the sole section his brother and mother had salvaged from E. J.'s estate.

I was born in Hibbing, Minnesota, during the blizzard of January 10, 1951, one of the worst to hit the Midwest since the notorious blizzard of 1888, which struck on the day of Tessie's birth, exactly sixty-three years before. In our parts children are conceived at the first thaw and born in blizzards.

The cycle of generations runs long in the Murphy family. I was Tessie's first grandchild. We were extremely close, and I was steeped in her tales of struggle. Most children told of the bad old days squirm and close their ears; but I listened, and those stories haunt me still.

The Blighted Tree

"Spare that sucker at the root,"
she cautioned as I felled her tree.
Half-blind and eighty-three,
she planned to watch that sprout
blooming and bearing fruit.

Hardy enough to outlast me,
humming with bees and memories,
this offshoot of a blighted tree
spreads its flowery boughs
outside Tessie's shuttered house.

After World War II, my father left the farm and worked his way through college. Earning a master's degree in speech at the University of Minnesota, he joined the Minneapolis faculty as an assistant professor. There he met a beauty from Duluth, an aspiring actress who took the leads in university theatrical productions. His fate was sealed when he saw her play Lady Macbeth. To this day he looks at her and mutters: "Screw your courage to the sticking point."

Mother's prospective career was derailed when she bore five children in six years. In retrospect her sacrifice and devotion seem so daunting to us all that none of us has emulated her. Like so many of our fellow Boomers, we have small families or none; and we put our careers first. Yet we sometimes wonder what we're missing.

Rather than try to support a large family on an academic's salary, Dad decided to go into business. He quit his job at Hibbing Junior College and signed on with Connecticut General, then a thriving and respectable life insurance company, not the spendthrift colossus it has become as Cigna Corp. After a stint at the C. G. office in Duluth, Dad moved us to his home territory—Moorhead, Minnesota, a town of thirty thousand just across the winding Red River from its larger neighbor, Fargo. It was a pleasant place, modestly prosperous, its businesses based on the juncture of roads and rails at the center of the Valley. Over the decades since E. J.'s day, homesteaders' elms had grown into an urban forest, breaking the prairie winds and hiding the emptiness of the horizon.

Here Dad opened an office of his own, where he showed farmers and business owners how to protect their heirs from the claims of creditors and the IRS. Each summer at our lake cottage, forty miles to the east, Mother staged performances of *A Midsummer Night's Dream* with her offspring, and in the winter she organized children's theater companies to tour tiny North Dakota towns. Though I inherit my passion for literature from both sides of the family, it was my mother who introduced me to Shakespeare at the age of seven.

The Sage Hen

To slake her fledglings' thirst
she dowsed her downy breast
and flew through blowing dust
from the river to her nest.

Now she is distressed,
always dreading the worst
for the flighty brood she nursed
because we do not nest.

After four years of college on the East Coast and several more as a trainee agent at C. G.'s Minneapolis office, I returned home and joined Dad's business. In 1978 we worked for Ivan Miller, who at age forty-eight was one of the wealthiest farmers in North Dakota. We persuaded him to gift his three sons large parcels of his rapidly appreciating land, and to discharge the incompetent bank trustee who would have managed his estate in the event of his death. He also assisted his sons in purchasing a substantial life insurance policy. Two years later he was struck by a loadout chute and smothered under tons of sunflowers.

There is a toughness and single-mindedness in farmers unlike any I've seen. It's not easy to persuade them of anything. During the winter before Ivan's death, I tried to get him to buy more insurance, arguing that the continued inflation of land values could raise the death taxes imposed on his heirs. At first he waved me off, saying: "If they tried to sell my land, they'd break the market and there wouldn't be any taxes." Later, when he'd looked over my calculations, he agreed to buy when the wheat was green. He didn't have that much time.

But the sons inherited the father's toughness along with his land. Once Ivan had purchased a new farm and sent his youngest son, Jon, who was only nine at the time, to watch the crew sent out to work it. The men, who were mostly Ivan's age, went to the wrong farm. An argument ensued between Jon and the crew. Ivan came by and was horrified to see four tractors sitting idle. He roundly chastised Jon for failing to manage the men.

Such an upbringing girded the young Millers for confronting the IRS. After a protracted wrangle, they emerged victorious with their farms intact. The eldest, Kelly, told us on a pheasant hunt: "When Dad died, the rumor was that we would have to sell half our land, but thanks to you two, we have a lot of disappointed neighbors."

The following spring Kelly invited me to join him in a partnership and purchase a farm on the higher ground west of the Valley, where land prices were lower. Thus began my friendship with a man whose crimes against the language conceal one of the shrewdest minds I've ever encountered. When wheat prices crashed in 1985, Kelly said: "The farmers are totally diswildered." Kelly cash-rented his first farm at age twelve, bought his first section at sixteen, and has experienced enough diswilderment to be a consummate pessimist.

Kelly's Lament

I fear for my spring wheat.
Will it grow red and tall
  or head out small?
Will it succumb to heat,
  drought and dust
  or rot and rust?
Will it be flooded out
or flattened by the hail?
  I am beset
  with doubt and debt.
Surely the wheat will fail.

In the dry and windy May of 1982, the soil blew away as tractors worked it. Tillage practices had changed little since the Dust Bowl. Too many farmers ploughed their fields to powder between crops. On the day Dad and I were to negotiate the purchase of my partnership's first farm, a wall of dust rolled over town. The cloud was electrically charged, and lightning flashed through the brown dusk at noon.

We had sought spring possession and the newly seeded crop. But the farmer demanded such a ridiculous premium for his fall tillage and spring seeding that we decided to wait out the drought and seek possession the next fall. I told the man I didn't want the crop at his price, and he angrily accused me of reneging on a verbal agreement. Dad said: "Mister, if you're going to impugn my son's integrity, we'll adjourn this meeting to the parking lot and watch your land blow by until you choke on your own grit." The farmer was so taken aback he lowered the price by $50,000, and the deal was done.

Our newly purchased land lay near Gwinner, North Dakota, gently rolling country some twenty miles west of the Valley's edge, at the primordial boundary between tall and shortgrass prairies. It was a chancier place to farm than the prime bottomland Kelly owned near Wahpeton. Some years the summer storms form over the rangelands beyond the Missouri and water our whole region. Other years the rains fall only to our east, over Minnesota lake country. The heavy soils of the Valley can hold enough moisture to sustain a crop through a dry spell; our western coteau with its lighter soil would not. Land prices reflected the difference. Our venture was a calculated risk.

By the time of our first harvest, I was a wreck, with a crick in my neck from scanning the sky. Kelly's burly foreman was unloading a combine as I drove into the yard. He strolled over and said with a stricken look: "Kelly's sendin' out the ploughs 'cause it don't pay to combine this junk." I nearly had a stroke, but our wheat made forty bushels while farms five miles away ran twenty.

Farming All Night

I dreamed of a lush stand of hard spring wheat
    and bumper barley yields
    ripening in my fields,
sunflowers blooming in the summer heat—

then came the black squalls with swaths of hail,
    lodged and battered grain,
    ruinous harvest rain
and flooded barley rotting in the swale.

Our second summer it scarcely stopped raining. There was a constant threat of severe weather. One swath of hail passed a mile south of our home half, smashing beans, trashing grain, and stripping corn to the stalks. Two weeks later another hailstorm totaled a fine stand of wheat right across the road from the luckiest section in the township. We didn't lose a spear.

As we walked the sodden fields in July, Kelly and I marveled at the heaviness of the heads, the thickness of the stand. In August it turned hot and dry, and the wheat was finally fit. Thirty-foot headers slashed through the straw, and the choppers screamed as they sheared and scattered the chaff. Kelly's brothers threw four of their combines into the fray, and seven green behemoths lumbered over the land—Deere 8820s, the pride of the Millers' fleet, unloading their hoppers on the run as semis roared in and out of the fields. I phoned Dad to tell him the wheat was going great but the trucks couldn't keep up. He laughed and said: "Every farmer should have your problem."

As my father knew, farming is an unpredictable and competitive business, despite all efforts to portray it as a bucolic "lifestyle." Nostalgia has its place, but a farmer who thinks he can live on it will not be farming long. The necessary concomitant of the right to succeed is the room to fail. An overpriced loan or an ill-timed delivery contract can be as devastating as a hailstorm. The marketplace prices everything, even loyalty.

Eight years after our first harvest at Gwinner, Ron Offut, owner of a dozen John Deere dealerships, visited Dave Miller's office, trying to sell fifteen combines to replace the family's aging fleet. For an account like that, Offut was ready to give a client plenty of personal attention. But over the conference line Deere's competitor, Case, made an offer Offut couldn't match. The next fall Big Red replaced Big Green on fifty-five thousand acres.

The Failure

Tractor and combine axle-deep in muck,
seedcorn and soybeans frozen in the field,
the home farm pledged against a bumper yield,
he has run out of money, time and luck.

What would his frugal Swedish forebears think
to see their hard-won holdings on the block?
There is no solace for a laughingstock
in woman's arms, religion or strong drink.

Any day now the banker will foreclose,
summon the sheriff and the auctioneers.
What will he tell his sons in twenty years?
He cannot wholly blame the early snows.

When crop prices and land values fell in the mid-1980s, we had to expand or get out of farming, so we enlisted capital contributions from three friends and our little venture grew.

Our boldest acquisition was the Femco Farm #1, once the crown jewel of an agricultural empire, with one of the largest contiguous chunks of land in the Valley. Publisher of the *Minneapolis Tribune,* Frank E. Murphy had owned a slew of farms. No relation to me, he was a lace curtain Irishman, whereas by the Thirties my kin had reverted to the potato bag variety.

Years before he was elected Governor of North Dakota, Bud Sinner (whose brother, Father Sinner, married my sister Mary to a man named Marion) told me an anecdote about Femco. There was a famous herd sire on that farm. He was a Grand Champion with a passel of blue ribbons, and the newspapers wrote about him so often that Bud's kids had nagged him all summer to show them the greatest bull in the Valley. One Sunday Bud loaded the nine of them and two dogs into a station wagon and wheeled into the Femco yard. As the children and dogs piled out of the car, Bud walked up to a gray-haired farmhand and announced: "I'd like to take my kids into the barn to see the bull." Incredulous, the old man asked: "Are all dese your kids, Mister?" Bud acknowledged it was so, and the old man said: "I bring the bull out to meet you."

Nowadays newspapers are more apt to tout the new corn plant (government-mandated ethanol for fuel) or the sugar beet mills (quota-protected domestic market) or the french-fry factory in Jamestown (site chosen after blackmail of town council for tax rebates). To stay on the land, farmers are compelled to cultivate big government and big business as carefully as we tend our fields. Ever more aware of our customers and competitors in the global market, we pray for lower interest rates, for peace in China, for drought in Ukraine.

One thing never changes. We hazard all each spring, maybe for the last time. Yet risk has its rewards, and for me there is no greater satisfaction than seeing fat partridge flush in front of the combines as the sun sinks into our wheat.

Harvest of Sorrows

When swift brown swallows
return to their burrows
and diamond willows
leaf in the hollows,
when barrows wallow
and brood sows farrow,
we sow the black furrows
behind our green harrows.

When willows yellow
in the windy hollows,
we butcher the barrows
and fallow the prairie.
The silo swallows
a harvest of sorrows;
the ploughshare buries
a farmer's worries.

Now harried sparrows
forage in furrows.
Lashing the willows,
the north wind bellows
while farmers borrow
on unborn barrows.
Tomorrow, tomorrow
the sows will farrow.

SNOW GEESE

1988

Feathers

I.

A warbler yellower
and smaller than a flower
trills
among the daffodils.

II.

Sipping a cordial from
July's geranium
a humming ruby throat
floats
flaunting its showy coat.

III.

Fog.
A pheasant hunter slogs
as his quartering dogs
romp
through a muddy cattail swamp.

IV.

A blizzard-blown bunting
no one would dream of hunting
spins
in the finger-drifting winds.

When I was still too small to carry a gun, Dad would pick me up after school; and wide-eyed I'd watch him bag his limit of pheasants within ten miles of Fargo. But the Soil Bank ended when I was nine. By the time I finished college, we drove a hundred miles and walked five to flush one rooster.

We farmers were part of the problem. In the Red River Valley, not 2 percent of the land was fit for a pheasant to hide in. Such prime bottomland is too precious for wildlife. On the coteaus it's a different story. Sloughs and potholes abound, better suited for raising ducks than wheat. Pastures and shelterbelts offer cover and forage for gamebirds. But economic pressure shapes the use of land, and government can only distort the realities of the marketplace, not repeal them. In 1983, before laws turned Draconian, Kelly and I scraped a shallow ditch through a new half section of upland. Draining three sloughs, it added twelve tillable acres and five hundred bushels of wheat. We needed every kernel to service our debt.

As a hunter, my interest ran the other way. Hoping to save or restore habitat, I gave generously to Ducks Unlimited and welcomed the Conservation Reserve Program, which idled millions of marginal acres in less productive regions at taxpayer expense. Upon enrollment farmers agreed to fallow their qualified lands for ten years and seed them with natural prairie cover. For once the government was doing something right. Even a blind old sow roots out an acorn now and then.

Of course Congress couldn't quit while it was ahead. Wetland and set-aside policies now encroach on the use of private property to such an extent that farmers seem to spend more time and money on compliance with bureaucratic diktats than they do raising their crops. No wonder the average age of a North Dakota farmer has risen from fifty-one to sixty-two since 1980. Ducks, geese, even pheasants are plentiful again; but farmers are a fading breed.

Next Year, Locusts

Plough the stubble. Set the drawbar deep.
    Let the north winds blow,
    smothering the fields in snow.
Farmers and the depleted soil must sleep

before the thistle thrusts a thorny shoot
    from its bristling corm,
    hatching locusts swarm
and the first cutworm chews a tender root.

What brother Jim and I know of hunting we learned from our father and his best friend Ed, a gruff, gravel-voiced old man who farmed all his life east of Moorhead. We learned to analyze flight and feeding patterns. We became good enough scouts that when October dawn flared over the flat horizon we were often in a field where snow geese were determined to feed. We learned to array decoys so they look realistic from a flock-leader's viewpoint and to sound our calls convincingly enough to deceive the younger geese. We learned to lead, swing and squeeze, and mark the cripple that sails afar.

When I shot my first goose, Dad and Ed and I were hunting in pits. We didn't own any decoys in 1964. Instead we weighted down forty white paper sandwich bags with clods of earth and turned them into the bitter wind. After what seemed an eternity, a small flock set their wings and swooped straight into our spread, which must have looked more like a White Castle parking lot than a flock of feeding geese.

I was seized by an acute attack of buck fever. The birds looked big as B-52s, and I blasted away when they were still fifty yards out of range. I was gently chastised and learned never to shoot until the men opened fire. A little later we heard a puzzled squawk. A naive fledgling had sailed into the paper bags without our noticing it. We stood up and shouted, but the foolish bird refused to fly. So I potted my first goose on the ground.

Jimmy's first goose was a cleaner kill. With neither Dad nor Ed to scout for us, we'd spotted a huge flock feeding at sunset just west of Starkweather. We set up in a rock pile an hour before sunrise and nailed a couple of mallards while the sun was a streak of gold in the East. Then ennui set in until ten o'clock or so. Finally a quartet of yearlings started to decoy, circling and recircling, getting tantalizingly close. At last they flared straight overhead, and we dropped two. The survivors, thinking that their siblings were gorging on the tender green shoots that carpeted the ploughing, swung round one more time and set their wings to die.

The Recruit

Memorial Day, 1997

An honor guard of battle-scarred old men
discharges antique carbines at the sky
as though the ghosts of war were winging by
like pintails flushing from an ice-rimmed fen.
How many of these troops will hunt next fall?
Fewer and fewer totter out to shoot.
They hardly hear the mallard's bugle call
which lures me to the sloughs with my recruit—
a boy shouldering arms where reeds grow tall
and mankind's present enmities are moot.

I hunt with a purity of purpose which I devote to few endeavors; yet I am only a fair hunter. Jimmy, on the other hand, is a peerless outdoorsman. He has canoed the nearly impassable rapids of Canada's Back River, clear to the Arctic Ocean. He is an accomplished hiker, camper, and skier. Though less experienced than I, he is a fine sailor. I'd never essay Lake Superior in a big sloop without him, for he is the equal of any three-man crew. But nobody instantly masters a difficult art, and Jimmy wasted as many shells as anyone. His quantum leap to proficiency came while we were ditch-blasting near the South Dakota border.

By the time the geese ran the gauntlets of Manitoba and North Dakota, Ed would say, "They don't decoy worth a damn, because the dumb ones are all dead." However, there is a magical day nearly every year when roughly 150,000 geese are concentrated on the Sand Lake and Jim River refuges. It is the day of freeze-up. The geese, sensing the inevitable and knowing how far away lie the Texas marshes, fly into a feeding frenzy. A vast, white wave, they rise into the dawn, their cacophonous calls and the rush of their wings audible miles away. It is snowing in big, dry flakes and blowing hard. The sky is thronged with geese who fling themselves into the cornfields one last time.

On one such day Jimmy shot his limit in an ecstatic burst. Forgetting the calculated vectors of bird speed, wind, and range, he swung his gun instinctively. In the parlance of Star Wars, the Force was with him. Blasting from ditches with his double-barreled twelve, he scored two doubles and a single at very respectable range.

Colorblind

for James Murphy

Hunters agree our labs are black
  and see the snow as white,
but lesser hunters simply lack
  the Murphy brothers' sight.

Why should greenheads not be red
  in the potholes' purple sheen
when the so-called red on a rooster's head
  glows like the sunrise, green?

Never would I have believed Ed's preposterous tales had I not personally witnessed so many of his exploits. One bluebird day it was deathly dull in the decoys. Geese flew far overhead, honking derisively at our setup. A flock finally alighted half a mile away, and Ed and I set off in his old Imperial to "sneak the geese." This is a haphazard procedure which rarely succeeds with such wily quarry, but at least we would rid our neighborhood of a thousand live decoys that constituted serious competition for our plastic frauds.

We drove round the section to a tree claim with slumping, derelict buildings. Beyond the trees we could see the geese feeding within range of the field road. Ed gunned the car, slammed on the brakes, and shot out his window as the flock took flight—a breach of etiquette and law forgivable only in a man who had to feed his family on game during the Depression. Had I attempted the feat, I would have wildly flock-shot and been lucky to bag a random goose. But Ed selected a pattern of three for his first shot, two for his second, then dropped the tardiest goose with his third. Six geese fell like stones at distances we paced off between sixty-five and eighty-five yards.

At four o'clock the next morning I stumbled sleepily into the motel cafe. Over their black coffee several young hunters were snickering about a ridiculous old man they'd met in the bar who claimed he shot six geese with three shells. I strode over to their table, swore I had witnessed the deed, and told them Ed was the best goose hunter in North Dakota. We left unspoken the thought that his grandmother must have slept with a Chippewa.

Though we knew them all by rote, my brother and I never interrupted Ed's hunting stories. They came from a different era, as Ed acknowledged himself when he bade us obey game laws to the letter. We would never experience a night like the one when his immortal dog Duke retrieved forty-nine ducks shot with forty-eight shells while a full moon shone on Lake of the Woods. Ed left us a world too populous and rule-bound for a born hunter with a gun dog's tireless thirst for the kill.

Duke figured in many of Ed's best stories. He would tell us how the boys at Sportland couldn't get Duke to budge from his "stay" position even though they waved a piece of tenderloin under his nose and Ed was nowhere to be seen. We shook our heads in admiration. "That must have been some dog."

The best Duke story I know to be true from a third party. Ed had a nearly fatal head-on collision in blinding snow. Duke emerged unscathed, but Ed was hospitalized in critical condition. Duke didn't eat, drink, or sleep for three days. At last Ed's son, fearing for the dog's life, phoned the hospital. As the receiver was held to the retriever's ear, Ed croaked, "I'm okay, Duke." The dog promptly devoured a bowl of pot roast and slept for two days.

In his later years Ed kept no retrievers, but we sometimes hunted with dog owners, and I envied them. One weekend in 1985, I tracked pheasants behind Jericho Jones, a magnificent black lab. Through the densest cattail sloughs where a man alone wouldn't have a prayer of flushing a pheasant, Jerry worked uncannily, never getting out of range, circling dodgy roosters and flushing them into the guns. Just as miraculously, he retrieved every bird we shot. For all its brilliant color, a fallen pheasant is invisible against any backdrop but black dirt. Yet in that jungle of ricegrass and cattails Jerry marked by sound and retrieved by scent. By Sunday afternoon I was sold.

Jerry's owner referred me to his trainer, Carl Altenbernd, whose Gun Dog Kennels are known across the Valley for superior stock. Carl has the most remarkable rapport with dogs I've ever witnessed. With a minimum of force he can reduce the wildest two-year-old male to submission in minutes. His loving encouragement coaxes *sit*, *stay*, and *come* from the dimmest of puppies equally fast. Carl had a small seven-month bitch named Dee. Eager and obedient, she marked and retrieved well for a puppy. After checking my references, Carl sold her to me. I registered her with the AKC as Diktynna Thea (Greek for Diana, Goddess of the Hunt), and I have yet to be shredded by Actaeon's hounds for my impiety.

Within weeks Dee ceased to sit, stay, heel, or retrieve. I returned with her to Carl, who watched me try to work her. Without another glance at me, he smartly put his protégé through her paces, then asked: "Dee, do you think he's trainable?" Thus began the exacting process of teaching me to collaborate with a beast whose hunting instincts are so acute all I need add is a 4x4 and a gun.

Diktynna Thea

for José Ortega y Gasset

We hunker by the fire
to read a hunter's praise
of Socrates. The blaze
leaps like a dog's desire
when ducks circle a blind,
then gutters and burns low.
Outside the moonlit snow
flows in a bitter wind.
I scratch my bitch's withers.
She sighs for whirring Huns,
cackling cocks, blasting guns
and a mouthful of feathers.

In *Meditations on Hunting* José Ortega y Gasset wrote: "Here is the dog, which is an enthusiastic hunter on his own initiative. Thanks to that, man integrates the dog's hunting into his own and so raises hunting to its most complex and perfect form. This achievement was to hunting what the discovery of polyphony was to music. In fact, with the addition of dogs to beaters and shooters, hunting acquires a certain kind of symphonic majesty."

For Dee's first duck hunt we joined Jericho and the Joneses, surrounding a shallow slough with six guns. The mallards divebombed the decoys, and drakes dropped from the sky as multiple volleys roared. Dee was totally "diswildered," and Jerry smartly fetched every duck to his master's side. Later Jerry's owner rose to count the ducks and shouted, "They're gone!" Dee had sneaked through the switch-grass, stolen all of Jerry's ducks and piled them neatly beside me. Aye, a braw retriever!

Later that fall Dee skittered and crashed across thin ice to retrieve a crippled mallard. Breaking through time after time, she hauled herself out of the frozen slough with her dew claws and persevered to her goal. For the return trip, her brain overruled her adrenaline, and she rounded the slough on dry ground.

The only time I ever saw Dee stop in her tracks was when she encountered Mexican sandburrs in Kansas. After I finished picking the devilish spikes from her paws, we drove to a hardware store and purchased two pairs of heavy children's mittens and a roll of electrical tape. With the tape I secured the mittens to her hocks. The farmers hosting us howled with laughter to see her first tentative, stumbling steps. But when she hit bird scent, she charged through the weeds as gamely as ever.

Occasionally, taken by a fit of waywardness, Dee would chase hens, flush roosters out of range, and otherwise disgrace me. I too disgraced her when I missed an easy shot. Dismayed by the look of disdain in her brown eyes, I honed my own skills along with hers. We may never have been as formidable a team as Ed and the legendary Duke, but we did our best to select for intelligence among ducks and pheasants.

A Dog Young and Old

I. Obedience

I am the Alpha male,
dispenser of her meat.
With drooping ears and tail
she trembles at my feet.
Leader of the pack,
I growl a wolfish note,
tumble her on her back
and bite her furry throat.

II. First Spring

Daily the flocks increase
as the floodwaters rise.
Celebratory cries
of homecoming geese
boggle my puppy's brain.
Out of the melting snows
she lifts her curious nose
to scent the impending rain.

III. Skunked Again

Flurries of hoarfrost fall
like silver maple leaves
as laggard snowgeese call
and my retriever weaves.
Hot on a bird, she speeds
into the frozen fen.
From the last clump of reeds
bursts an indignant hen.

IV. Dog Heaven

Sprawled in the pickup box,
my old bitch is half-dead.
Her slashed teat needs stitches,
her nettled nose twitches,
and nine gutted pheasant cocks
pillow her dreaming head.

Ed never lived to see Dee in her prime. A lifelong smoker, he developed lung cancer in 1984. A few days after his first surgery he suffered a massive heart attack in which the entire upper chamber of the right ventricle blew out. The surgeons spent eight hours stitching it together with Teflon-coated wire. Dad waited all night with the family. Periodically the chief surgeon emerged bespattered with gore to suck down a cigarette of his own and remark that Ed's chances were zero. But two days later he woke up and demanded a transfusion of Chippewa blood. Shocked, his nurse asked: "Whatever for?" And Ed growled: "Because I'm a hunter."

Ed recovered after a fashion, and we took him hunting again, though he ruefully observed: "I couldn't hit a bull in the ass with a pail full of sand." The next year cancer invaded his liver, another organ that had seen its share of abuse. Again the doctors told Ed his time was up, but he wasn't ready to quit. Another hunting season was approaching, so he ordered yet another brutal round of chemotherapy and outfoxed the doctors again.

Come opening day, we were set up four miles south of Snowflake, Manitoba. A howling Alberta Clipper brought six inches of snow, and the geese flew over our decoys at slingshot range. So Ed's last hunt was a glorious enterprise.

When my father delivered the eulogy, he quoted Carl Sandburg. "*To live hard, work hard, and die hard and then go to hell after all that would be too damn hard.* Ed liked those lines. They were reminiscent of his early farm days when everything he lifted was heavy and everything he bumped was hard."

Now the wheat is in the bin. The beans are being combined, and the corn is not far behind. It is six o'clock in the morning, and dawn is a gray smudge in the East.

The Blind

Gunners a decade dead
wing through my father's mind
as he limps out to the blind
bundled against the wind.

By some ancestral code
fathers and sons don't break,
we each carry a load
of which we cannot speak.

Here we commit our dead
to the unyielding land
where broken windmills creak
and stricken ganders cry.

Father, the dog, and I
are learning how to die
with our feet stuck in the muck
and our eyes trained on the sky.

TUMBLEWEED

1966

Spring Song

Oh the grouse are loudly drumming
and raucous geese are crying.
Summertime is coming,
flocks of ducks are flying,
Lovesick lads are sighing
and bumblebees are humming.

Oh the hummingbird is sucking
nectar from a lily
and the yearling colt is bucking
an unwilling filly.
Squirrels are scolding shrilly
and sparrows wildly fucking.

A child rejoices in spring for the nearness of summer and freedom from school. Only a hunter could welcome the return of spring because it heralds the approach of fall. Yet the farmer in me dreads April as the cruelest month because it begins another season of sleepless paranoia over the fate of my crops. Thus have hunting and farming skewed my attitude toward the seasons. Only in winter are the land and I at rest. So it was with mixed feelings of anxiety and anticipation that I used to make a spring pilgrimage to Teewaukon Refuge, where I greeted the returning geese and inspected my nearby fields newly emerged from the last gritty drifts of snow.

As I passed through Gwinner and the first meadowlarks sang on fenceposts, I was always more eager to see our worst field than our best. Though none of our land was sufficiently erodible to be enrolled in the Reserve, we had one poor section that was a wildlife paradise. With four miles of shelterbelts and twenty slough bottoms, it harbored a small flock of pheasants when we bought it. We planted feedplots of corn and sunflowers to tide the birds through blizzards. We scattered crushed oyster shells for hens to peck and improve the viability of their eggs. We summer-fallowed parts of the farm and minimized use of pesticides. The result was a tenfold increase in pheasant population.

We undertook these improvements without government support. Farmers aren't enemies of the land, though we're sometimes driven by short-term necessity to act as though we were. Then we have to live with the consequences, just as surely as hunters who kill off hens shoot no pheasants the next year. A civil society tries to set rules for the common good, and we abide by them as best we can. But consensus is preferable to mandate. Congress has passed no law requiring farmers to change tillage practices and minimize wind erosion, yet we've rarely seen blowing dust in the Valley since the early 1980s, when no-till drilling caught on. If we do get dust these days, it's coming off quota-subsidized beet fields, which are still ploughed to powder every fall.

Lost Causes

I. Betting the Ranch

He could have sold his pregnant cows last fall.
  A hedger ought to weigh
  the cost of air-dropped hay
before a blizzard and a margin call.

II. "Blow, Winds, and Crack Your Cheeks!"

A stage so large the combine seems a prop:
  the farmer plays by heart
  Tom o' Bedlam's part,
and hail drops like a curtain on the crop.

P. J. O'Rourke has pointed out that during Pennsylvania's Whiskey Rebellion the policy of the federal government was to shoot farmers. Today we are abject subjects of the U.S. Fish and Wildlife Service, the Occupational Safety and Health Administration, the Environmental Protection Agency, the Agricultural Stabilization and Conservation Service (lately merged into the Consolidated Farm Services Administration), the Army Corps of Engineers, the Commodity Credit Corporation, *et cetera ad nauseam*. Shooting us might have been kinder and gentler than this slow strangulation.

Picture three hundred thousand United States Department of Agriculture bureaucrats regulating one hundred thousand serious commercial farmers. My apple orchard on the outskirts of Fargo, a hobby farm that covers six acres and grosses two thousand dollars in a good year, would entitle me to all the offices of government if I owned no other land. Virtually every county in the United States, from desert to suburb, has its own USDA headquarters. Recently in Fairfield, Connecticut, the Department disbursed just one check—fifteen thousand dollars to the local hunt club for a manure spreader. Horseshit! An alarmingly plausible story has it that one county agent was found sobbing at his desk. Asked why, he replied, "My farmer died."

Created in the name of preserving the "family" farm, the Program has grown into a million-teated milch cow for fat cats, who suck up almost all its funds and lobby relentlessly for more. Leftists propose means-testing as a remedy. Right-wingers respond with the "Freedom to Farm Act," which barely begins to weed out the abuses. Far better to phase out the whole Program in a final Five Year Plan. Have we learned nothing from the examples of Russia and China? It's a true testament to the resiliency of Americans that we've survived so many Five Year Plans without turning in our ploughs.

Handsome Dan

Like a horny boar
sniffing a sow,
he was eager to lend
on hardscrabble ground.
Now the hapless man
who shook the hand
of handsome Dan
loses his family's land.
Like a squealing sow
on the slaughterhouse floor
he forfeits his life
to the banker's knife.

With reliable partners and bountiful crops, I expanded my farm operation to five thousand acres by the late 1980s. In that decade of easy credit, bankers were always ready to lead a man into deep water and ditch him there. I still can't believe they let me speculate on the futures markets with borrowed money. Just like my grandfather, I was in way over my head, but I didn't know it till the drought hit.

From 1984 to 1987 crops had thrived in the Corn Belt, but interest rates were high and commodity prices abysmal. During this severe agricultural recession, a fourth of the Midwest's farmers were forced off their land. I had thought to take advantage of a bottom in land prices, but I found myself owing more than my farms were worth. Then came 1988. Cash grains soared to levels not seen since 1973, and I had nothing to sell.

The rains failed in 1987, but the soil had retained moisture from a wet summer the previous year, and the crops survived some hot days to bring a fair yield. After harvest the weather stayed abnormally mild and sunny. By Christmas we still hadn't seen snow, and I knew we were in trouble. In January an ice storm ripped limbs off trees all over the Valley, an event unheard-of at midwinter, which is often too cold even for snow. That was the last meaningful moisture we got till September.

Spring brought blistering sun and howling winds. My orchard bloomed three weeks early, but a dust storm ripped every petal off the trees, and the temperature fell from 82 to 23 in twelve hours. A week later it was 94. With some trepidation I okayed a scheduled addition to a shelterbelt. Of two thousand spruce seedlings, all but seven were dead within weeks.

My Gwinner farms caught some spring showers, so at least the crops germinated. By July the upland wheat had burned to straw, and I pinned my dwindling hopes on beans. Late in the month an isolated thunderstorm parked over my fields one evening and dropped three inches of rain. I watched the lone cloud towering in the distance and crossed my fingers. When I drove out there the next day, the soy and navy beans had already burst into bloom. Two days later it was 102, and the blossoms wilted. Though the plants survived to the end of the season, they never set any pods.

All told we endured thirty-nine days over 90 degrees, beating the record set in the Dust Bowl summer of 1936. Rainfall for the calendar year totaled barely eight inches, about the same as Phoenix, Arizona. The only things that thrived were weeds and grasshoppers. Milkweed, cocklebur, ragweed, quackgrass, black-eyed susans—we had them all. But beans were fetching $10 a bushel. Surely the rain would come again and the heat relent. So we sprayed Basagram and Blazer on the weeds, throwing good money after bad.

In the heavy loam of the Valley, our wheat fared a bit better, until the locusts started to swarm. Our Femco farm was hopelessly infested. First we sprayed diazanon, then malathion, neither of which worked. At last we resorted to methyl parathion. At the cost of killing every songbird on the farm we saved a handsome stand of wheat. Just as the heads were filling, we hit the summer's peak heat of 105 degrees. The plump heads dried into empty husks, and the wheat made fifteen bushels. We were lucky. Kelly's home farm five miles away yielded only six. On one 750-acre field he ran five combines for eight hours to fill a single truck. Kelly's brothers, who were among the few to get half a crop, ran their trucks at night to conceal their good fortune.

Gallows humor abounded. A farmer who normally harvests twenty-five thousand bushels told me he drove one combine day and night until he finished, then parked the machine in his shed and took a loan out on the two hundred bushels in his hopper. Another farmer claimed he saw two trees fight over a male dog.

Eighty-Eight at Midnight

A black calf bleats
at shriveled teats.
Incessant heat
withers the wheat
and wilts the silking corn.
Too few, too late
the spotty showers
mock my stunted flowers.
Too late I shrink from debt.
Like a spitted calf I turn
over a bed of coals
while the pastures burn.

West of us it was worse. I drove to the Beartooth Mountains on June 30 to catch trout, hike treeline meadows, and forget about the drought. First I had to cross six hundred miles of desert. Past the state capitol at Bismarck, Governor Sinner, in a moment of levity during better times, had erected a billboard which proclaimed: "Welcome to North Dakota. Mountain removal project completed." It should have read: "Abandon all hope, ye who enter here." The crops were shot, and mature shelterbelts were dying, half a century after the great tree-planting spurred by a previous Dust Bowl.

A south gale was blowing, and the temperature was 99 as I drove through Miles City, Montana. Upwind the pine-clad hills were on fire, and a pall of smoke mixed with the gritty dust of Yellowstone Valley fields. Dryland corn on deep alluvial soil was stunted and dead white. Even the irrigated sections were wilting in 5-percent humidity. Russian thistles, forced to ripen months early, were tumbling over fallowed fields like wind-driven drills, sowing their hated seeds. Even the sage had lost its spring foliage, and the ranges were scorched and bare of cattle. Whole herds had been shipped to greener pastures or slaughtered.

In Yellowstone Park I was appalled at the degradation of the high-country timber. It had been three years since I had passed that way, and everywhere stood dead or dying forest. "What this place needs is a good fire," I observed. Little did I suspect that within two months half the park would burn, sending plumes of smoke to cloud our prairie skies hundreds of miles away, where we were still waiting for rain.

The drought of '88 seared my soul. I learned firsthand that Nature can utterly betray us. I prayed for the Great White Combine to obliterate my dying but hail-insured crops. I lived the nightmare that had troubled my father's sleep for fifty years.

The Expulsion

Six weeks of drought,
the corn undone
and wheat burned out
by the brazen sun:

over that land
an angel stands
with an iron brand
singeing his hands.

Amid all the death and desolation of '88, a birth brought me comfort. Having acquired Dee at seven months, I'd missed her early puppyhood. By the time she was three, I felt ready to raise and train a puppy of my own, so I bred her with Jericho Jones. Carl Altenbernd helped me chaperone the big date at Gun Dog Kennels. Dee was very skittish, running around the shed, hiding under a trailer. At last she stood to stud, and the tie lasted almost an hour, as though Jerry were determined to prolong the chance of a lifetime. A kennel dog who never had a chance to chase bitches in heat, Jerry seemed to develop a whole new regard for me after that.

Two months later Dee began her nesting routine at 1 A.M. with a thorough tongue-wash of her whelping box. Knowing she was due, I was sleeping on a cot in the basement. Labor commenced around 4 A.M. Another hour and the first pearly amniotic sac emerged. Within squirmed a future hunter. By 11 A.M. Dee had eight sucking puppies. Jericho's owner got a pick of the litter. Luckily he didn't pick the little white-blazed bitch I'd already set my heart on. The other six pups went to hunters, farmers, and ranchers in three states.

My chosen puppy was so aggressive that I named her Maud Gonne, after the Irish revolutionary. Maud was the smallest of the litter; but since she was always the first on a teat at feeding time, she soon began to outgrow most of her siblings. By her fifth week I could throw a pigeon wing six feet in front of her and she would pounce on it, shake it furiously, then bear it back in triumph, as though she'd killed a wingshot goose.

When she got her growth, Maud became a tremendous predator. Her favorite pastimes were running deer through my orchard and treeing young truants who trespassed for fruit. Because she was born in the dry years, Maud never matched her mother as a swimmer. But she was that rare beast, a pointing Labrador. Upland game became her forte. Her early hunting seasons coincided with the introduction of the Conservation Reserve and great hatches of pheasant and partridge. With her long legs and rangy body, Maud was perfectly built for crashing through brush.

Passel o' Pups

Bonny bairnies, black an' fine,
wi yir yivver souks an' ruggs,
will ye be guid hountin dugs
worthy o' yir faither's line?
Will ye busk an' tak them doun,
frantik pairtrick, crouchin grouse,
an' the Deil's ain phaisant louse?
Will ye ding the raibbit broon?
Lak the dun deir ye maun lepe
owre yon scraggy, stany hill
whaur the wund blaws lood an' shrill,
sare an' snell, whaur muckle depe
drifts the snaw. Drink yir fill,
glazie beasties; souk an' slepe.

(see end notes for translation)

After her litter Dee aged precipitously. Always a quiet and undemanding animal, she took to lying in the orchard and munching on apples. Deer could pass right before her nose without raising her hackles. Only in the fall did she rouse from her torpor for the hunt, though she developed selective hearing, wandering off to flush rabbits or foxes, heedless of my whistle until it suited her to return. I came to rely more and more on Maud, who had learned to mark and retrieve better than any dog I've known. Tireless and high-spirited, she took after Jericho, now long gone.

Dee was only nine when she died one bitter winter morning. In her old age I often let her stay overnight in the house if the weather was harsh. Maud would come along, since I couldn't bring myself to abandon her to the elements while her mother enjoyed the comfort of a bedroom chair. My kennel dogs were slowly becoming house dogs. I heard both of them stir when I rose in the dark at seven to make coffee; but they would always lie low after I left the room, hoping to prolong their stay, no matter how badly they might need to go pee in the snow. When I returned an hour later, Dee was dead of her chronic apnea—the frightening pauses in breath which would follow seizures of twitching and snorting while she slept. This time she would never wake. Her outstretched tongue was already black. Maud lay silently at the foot of the chair, all the light gone from her eyes.

When the orchard bloomed in the spring, I interred Dee's ashes under the four Centennial crabapples which tower above neighboring Jonathans and Mackinacs. It was a warm, sunny day with a southerly breeze, and the petals flurried down like snow over Dee's grave.

Air

Come Diktynna come—
once more the mourning dove
coos in the blooming plum.

Mark the wood duck drake
plummeting through a grove
and greenheads in the brake.

When fledglings try their wings
and waves of migrants heave
skyward in widening rings,

you will not heed my gun
or leave this grassy grave—
your hunting days are done.

Drowse Diktynna drowse
lulled by a humming hive
under the apple boughs.

So stealthily stole death
my love could not retrieve
your evanescent breath.

WINTER APPLES

1996

Orchard in Bloom

These seven hundred trees
pillar a pagan church,
a basilica for bees
where seraphic songbirds
promiscuously perch
to blend their melodies
in canticles whose words
the vernal believer
need never decipher.

My orchard was planted in 1960 by Dr. Lee Christopherson, a Fargo neurosurgeon who wanted to see what apple varieties would endure our boreal winters. For his experiment he chose an old pasture in the low ground of a Red River oxbow seven miles south of town. With eighteen types of tree interspersed to minimize the spread of disease, his orchard developed a crazy-quilt appearance over the years, an effect only amplified when he built a large house across the road, designed to his specification with a monstrous array of solar panels in two tiers across the front. By 1985 Lee had fallen way behind on maintenance. His sickle bar had cut off the tip of his ring finger; his wheezing antique tractor kept breaking down; the mowing was never done, and the unpruned trees were running wild. He was relieved to find a buyer who wanted to restore the orchard rather than hack it down for a subdivision.

North Dakota apple trees are grafted at the root to crabapple stock, crab roots having more resistance than apple to the deep frosts of a long winter. Every year more of the old trees collapse under the weight of their fruit or succumb to the ever-present blight; so in spring, when the sap begins to run and the geese are calling, I graft Redwell or Fireside scion wood to the crab shoots that sprout from surviving roots. I'll never have the steady hand of the old doctor, or his remarkable ratio of grafting success, but I've stayed to some degree the decline of the aging trees. And I welcome an excuse to be outdoors in the orchard each April, to watch migrating bald eagles fish the river or kestrels hunt the hungry mice emerging from their winter nests.

Each fall I open the orchard to pickers before the start of dove season. Grandparents, parents, and children debark from their vans on sunny September afternoons to bag the ripened fruit and leave a few dollars in the mailbox. People still trust each other here. I've rarely caught anyone abusing the honor system. But the orchard remains strictly a hobby farm. Its income barely pays the cost of mowing.

"It Is Very Far North . . ."

Four giddy days are all that spring allows
the drunken bumblings of our honey bees
before a south wind, stripping petaled boughs,
turns apples into ordinary trees.
Ours have weathered blizzards, freezing rain,
a record flood crest and a May snow squall.
Now only scab, inchworms and hail remain
to rob us of an ample apple fall,
a brief lifting of limbs before the snow
grips them with such reluctance to let go.

The summer storms that water my crops often wreak havoc in the orchard. Sometimes the chainsaw whines for days, cutting broken limbs for firewood. Come fall, I have the sweetest-smelling hearth in the Valley. While I stack the cut wood behind the house, my young employees pile the slash with heaps of spring prunings in the orchard. As the weather dries and days shorten in September, we burn the piles on calm evenings, a country custom we'll have to quit when Fargo annexes our changing township.

Year after year people keep leaving the impoverished counties of the western Dakotas and seeking work in this growing city. As the glare of development edges southward, only the brightest stars still shine over the orchard at night. I used to see northern lights often when I first moved to the orchard in 1988. To glimpse them now, I'd have to drive far out into farm country.

Watching backhoes strip the richest loam in the world to make way for new roads and housing, I feel glad for the farmers who've reaped this ultimate profit from their land, but sad for their sons. Development affords the people of Fargo and Moorhead a prosperity they could never wrest from the soil. It also absorbs their children into a national culture which seems increasingly to erase all sense of place and heritage. I hope the boys I employ in the orchard learn to value hard work and the natural world more highly than their peers. At least they'll know the difference between serious issues (genetic diversity in grains) and spurious (Alar on apples) as an ever more populous planet tries to feed itself in the new millennium.

Like wheat rust or orchard blight, Dutch elm disease also demonstrates the fragility of monoculture. Valley towns support aggressive programs of pruning and spraying shade trees to buy time for the growth of more varied plantings. Such intensive measures are unsustainable for rural landowners. The bacillus claimed all the wild elms along my river oxbow in a single, disastrous sweep just after I bought the land in 1985. With so many breaks in the forest canopy, the surviving oaks, ashes, and lindens became as vulnerable to storm damage as the oldest, frailest apple trees. A few years ago the woods behind my house were shattered by two consecutive nights of hurricane-force wind as squall lines swept from Montana to Lake Superior.

The Track of a Storm

Bastille Day, 1995

We grieve for the twelve trees we lost last night,
pillars of our community, old friends
and confidants dismembered in our sight,
stripped of their crowns by the unruly winds.
There were no baskets to receive their heads,
no women knitting by the guillotines,
only two sleepers rousted from their beds
by fusillades of hailstones on the screens.
Her nest shattered, her battered hatchlings drowned,
a stunned and silent junko watches me
chainsawing limbs from corpses of the downed,
clearing the understory of debris
while supple saplings that survived the blast
lay claim to light and liberty at last.

A lifelong lover of trees, Grandma Tessie would have admired the last words a Scottish clan leader, The Douglas, spoke from his deathbed, instructing his eldest son on the disposition of his estate, "And if ye do naught else, lad, plant mony a tree."

Tessie never lived to see the orchard or the thousands of maples, ashes, and dogwoods I've planted in belts on a nearby field. Had she witnessed my rash acquisitions of land, she might have warned me against repeating E. J.'s mistakes. But Grandmother ended her days in Felton with her paraplegic daughter, Vivian. Nearly blind in her last years, Tessie welcomed a gift subscription to the large-type *New York Times,* which gave her a chance to relish Nixon's downfall and rail at Reagan's rise. While E. J. was alive, he and Tessie had canceled each other's votes at the polls. By 1980 it was my turn.

Though I saw my aunt and grandmother often, I spared them too much knowledge of my life. Their world was defined by Mother Church and the Democratic-Farmer-Labor Party of Minnesota. They never understood my departure from their faith, or my deepening conviction that the party of Hubert Humphrey no longer served the true interests of those it purported to champion.

When I was a boy, Tessie used to bounce me on her knee with the admonition: "Never forget that the farmer and the working man are always right." By the time Vivian followed her mother into the hereafter, the self-styled advocates of the working man had fashioned a cradle-to-grave welfare state. Why did I doubt this was what Jefferson had in mind for the party he founded?

Tessie's Time

She said the sundial stood so long
because it only counted hours
   when the sun was shining.

Its daily lesson kept her strong,
showing her how to husband powers
   despite their slow declining.

When the years totaled ninety-one,
   she was thirty-nine by the sun.

In the great work of Norwegian-American literature, *Giants in the Earth*, Per Hansa's wife upbraids her husband for building a new barn while they lived in a sod hut. She was consumed with envy for the new frame house built by a flashy Irishman, a newcomer to their South Dakota settlement. Per Hansa retorted: "A fine house never paid for a good barn, but a good barn has paid for many a fine house."

Even as our grain farms sank under the weight of debt and drought, our hog business burgeoned. Our original Gwinner land had included a small hog operation, which looked attractive as a tax write-off. We found a tenant and kept the business going, though it made no money during the first five years. In 1988 the tenant quit, but first he fulfilled his promise to locate a successor.

I owe that man a bushel of gratitude for getting me in touch with Rich Bell. After many years as an employee on corporate hog farms, Rich had gone into business with his sons in 1980, starting production near Breckenridge, Minnesota. By the end of the decade, he was ready to expand. In short order he filled our Gwinner barns with a herd whose advanced British genetics gave us a cutting edge with meat packers.

Rich had an idea for weaning a steady flow of profits from the notoriously cyclical hog market. He persuaded packers to give us long-term contracts indexed to fluctuations in the price of the corn and soy meal we feed our pigs. Thus we insulated ourselves from the major variable in the cost of swine production. With market risk controlled and income flows predictable, my partners and I committed millions in venture capital to the expansion of Bell Farms. We contracted with small producers in Iowa, Nebraska, and Minnesota, who knew they had to "get big or get out."

The Peg-leg Pig

for Rich Bell

A farmer's daughter keeps a hog
  who sports a wooden leg.
"Tell me about that peg-leg pig,"
  traveling salesmen beg.

"He saved me from a rabid skunk.
  He stomped it with his peg."
Suspiciously a seed man squints:
  "How did he lose the leg?"

"He found me when a whiteout hit
  and led me through the snow."
"You called the vet to amputate?
  A case of frostbite?" "*No.*

"He pulled me from a flaming barn
  before the rafters fell."
"Enough to put me off my corn.
  It must have hurt like hell."

"Who said my peg-leg pig was lamed?
  He never got a scratch."
"That leg is missing all the same.
  Sister, what's the catch?

"Was it chomped on by a bigger pig
  or torn off by a plough,
squashed beneath a threshing rig
  or trampled by a cow?

"Was the porker born to walk on wood
  or crippled in his prime?"
"Mister, you eat a pig this good
  one leg at a time."

From its humble start our partnership was soon competing head to head with Tyson Foods, Murphy Farms (no relation), and other giants of agribusiness. Just as the Bells boosted me back toward solvency, the final disaster befell my grain farms.

In the summer of 1991, while my crops were ripening on eight square miles of mortgaged land, a volcano blew its top halfway around the world. For a time Mount Pinatubo silenced the doom-mongers of global warming with an ash-pall of global chill. The summer of 1992 was the coolest on record in the upper Midwest. Daytime readings rarely hit 80 and nights were often in the 40s. The corn grew tall and green, but the ears failed to fill. Beans bloomed, but the pods never set. The ripening sunflowers were scarcely distinguishable from black-eyed susans. Only the wheat prospered, and that wasn't enough to pay the ever-mounting burden of debt. For the first time I was forced to sell land.

The next summer was worse: persistently cool, wet, and cloudy, though we were spared the Biblical floods that swept the Mississippi and Missouri Rivers. Instead the dampness bred an uncontrollable epidemic of fungal diseases in our crops. Rust and scab tainted our wheat with vomitoxins. The fields were impassable quagmires, water standing in the rows all summer. Mosquitoes hovered in swarms like banks of fog over the flooded ditches. Some farmers never even ran their combines that fall; they just waited for the first freeze and burned their blackened stands of grain. By the end of the season I had more land on the market, and no buyers in sight. The snows came early and deep, immuring stranded tractors in the unfinished ploughing. We didn't see dirt again for five months.

Failures of Promise

A flock of crows
found a road-killed ewe
frozen in the snow.
A drowsy bear
dragged a leg-shot deer
to its deadfall lair.
The lamb in the ewe
and the fawn in the doe
were devoured unborn,
and November snows
buried the standing corn.

I had suffered everything Nature could throw at a farmer. As unseasonal rains continued to pelt us in 1994, I sold the Femco farm for a modest gain, retaining in partnership with my brother only a few fields near Gwinner and a stake in our one big success, the Bells' hog operation. If farming has a future on these plains, it lies in value-added products like pork, or french fries frozen in Jamestown's new plant, or pasta produced from local durum at Cando. The land and climate are too unforgiving for a dryland farmer, even if he has another source of income to supplement crops that fail two years out of five. That our forebears thought such a place paradise is a measure of the hell they must have experienced elsewhere.

W. H. Auden once wrote: "I cannot see a plain without a shudder: Oh, God, please, please don't ever make me live there." As an undergraduate I was briefly fond of those lines. Disparaging the prairie made me feel sophisticated. Now I become claustrophobic when I visit Minneapolis, and I can't even face Los Angeles or New York, though friends or family dwell in each. I've lived here too long to feel comfortable anywhere else. Hellish as it sometimes seems, it's my native patch of hell. Some years ago a former classmate of mine observed in a letter: "All my friends have settled in mundane places like London, Rome or Singapore. Only you have chosen a truly exotic place to live."

I have done so because the changing seasons lend some rhythm to a disordered life, because here are my "roots," which we Dakotans rhyme with "puts." Raising wheat and hunting geese have wed me to the earth, from which my years out East nearly estranged me. Now I feel like that apocryphal old farmer who never left the state until his children took him to Colorado. Seeing the Front Range for the first time, he said, "These are rocks. In the north forty we have rocks. Take me home."

Buffalo Commons

In Antler, Reeder,
Ryder and Streeter,
stray dogs bristle
when strangers pass.

In Brocket, Braddock,
Maddock and Wheelock,
dry winds whistle
through broken glass.

The steeples are toppled
and the land unpeopled,
reclaimed by thistle
and buffalo grass.

GHOST FARM

1992

Author's Note: Shortly after he saw an early draft of this book, my father wrote his own recollections of boyhood in the Red River Valley. I include them here with his permission.

The unique thing about my childhood was that when I was born in 1917 my father was sixty years old and my mother only half that.

Ed Murphy had left Potsdam, New York, as a young man and worked in Iowa and Wisconsin to get together a stake so he could set some roots down. He was working on the Jim Hill railroad when he saw the Red River Valley and picked the area near Felton, Minnesota, as a future homestead site.

When he made up his mind, he smuggled enough lumber for an 8 x 10 shack and paid for shipping an ox and a horse, which arrived in 1878. He designed a way of hitching the two together and plowed his first soil on a quarter section about five miles northwest of Felton.

My father never talked much about the ensuing years, but they must have been difficult. He married a girl he had met in Iowa, and they raised eight children until she died in the early 1900s. Two of the children went to Iowa to live and the oldest daughter, Annie, became the mother to the nest.

In the next ten years, my father must have become a successful farmer and businessman. After starting out with a horse, an ox, and a quarter section of land, he owned three thousand acres, had thirty-six horses, two complete threshing units, a steam rig, and a gasoline outfit.

He was president of the school board when my mother came to Felton to apply for a teaching position. She got the job and he got the redhead two years later. During that time he really romanced her. She would go to summer session at Moorhead Normal School, thirty miles away by road, and he kept a horse and carriage so he could call for her at the dorm and take her riding around Fargo-Moorhead. He had a stable of riding horses and one named Zip Bang won at the North Dakota State Fair.

Despite the age difference, my father's persistence paid off and the marriage took place in 1915. Part of the plan was to live in the larger town of Ada so his four teenagers could go to high school. That is where I was born.

There is always confusion over what one remembers and what one was told. However, I vividly recall at a very young age being with my father and going everywhere. He was so proud of me and wanted to show me to everyone. When Molly came along and I was fifty, I could understand why, and I'm afraid I was just as big a show-off as my father.

I have pleasant memories of my older brothers. They teased me and played with me and really made too much of me. When Vivian and Dan came along, I faded from the limelight. By then I was in school. I remember my first teacher, Miss Holt, and recall being dressed up in pongee shirts, knickers, and knee socks. Soon we had to move to Felton, and going to school there was quite different. The boys wore overalls and were a tougher lot. I think I was still dressed up because it took awhile to become one of them.

My father bought an old square house on the road between Felton and Ada because the next year the main highline was coming through to give us electricity. The house was quite a contrast to Ada. We had kerosene lights and my mother cooked on a wood stove. The place was cold and had no indoor plumbing. But it was my mother who insisted we move. We were sliding backward economically trying to live in style in the big town of Ada.

I have a vivid memory of plowing in a big field. I was seven. My father had a huge Avery tractor which pulled six bottoms. There was also a three-wheeled tractor that pulled three bottoms. Then a triple plow pulled by eight horses and a gang (two-bottom) pulled by five. I brought up the rear with a two-bottom plow and five more horses. I had four reins to hold, two for the lead pair and two for the other three. I was to sit on the plow and stop at the end of the row to wait for the hired man. He would turn the outfit around and head me for the other end half a mile away.

All went well until I decided I could do this for myself. I got the reins tangled and turned too short and the plow tipped over. I jumped off unhurt and the horses stopped. My friend, the hired man, got it all straightened out and I don't think my father ever learned of it.

I have a keen memory of sitting on that plow watching the stubble go under, listening to the clink of chain tugs on the steel singletrees, smelling the sweat of the horses and harnesses. I thought of what a beautiful world I was growing up in.

One evening as I drove with the horses still hitched to the plow, they started to run the two miles home. At first it was a trot but then they were galloping full speed. The plow had only one wheel behind and at this pace it swayed from side to side. I hung on for dear life as they ran into the yard and straight into the barn, plow and all. My hired man friend said, "Well, you certainly were in a hurry to get home tonight!"

Threshing was a distinctive time, the reward for the spring's and summer's work. Sometimes there was too much rain. One year my father had the bundles hauled out of the water via a "stone boat," which was a rude sled drawn by a horse. More often it was too dry. The grain was short with empty heads. The kernels were small and went through the threshing machine into the straw stack. The straw was sometimes diseased and the whole countryside would be lit up with burning straw stacks.

During the first few years after the move to Felton, the crops were pretty good. This made for a joyful harvest. Neighbors joined together to make a crew and the machines would move from farm to farm. The ladies would cook enormous quantities of food—loaves of fresh bread, meat and potatoes for all three main meals, lunch at 10 A.M. and at 4 P.M., supper at 9 P.M. Afterward there were cakes and pies and doughnuts, and always gallons of coffee.

I think I was seven when I first joined the threshing crew. The steam outfit was gone because it was too cumbersome and the demand had diminished because more farmers had bought their own machines. I remember one day when my job was to load the grain tank. The grain was elevated to a bucket holding a bushel which then tripped and released the grain into the tank. I had to shovel the grain around and load the tank evenly. I was about through with leveling off 125 bushels when my father appeared and said, "Vincent, I want you to drive this tank to Borup to the elevator."

Borup was seven miles away. I couldn't believe he wanted me to go but I was so proud that I drove away feigning confidence. I had our two biggest horses and I recall how they strained getting the wagon out of the soft field. Once we hit the highway things got easier but it was a long and lonely road. In the elevator the horses were unhitched and a hoist raised the tank so the grain would run out. My vision from the loaded wagon had been fine but on the way back I was standing on the bottom of the tank and I could barely see over the top. It was starting to get dark and I was some frightened. But the horses knew what to do and my father was waiting for me with great praise. I grew some that day.

I never hunted with my father. I know he loved the sport and was a good shot. He bought a Parker Brothers shotgun when no one in our area had even heard of them. My brother Leo let me shoot the gun when I was about nine. I fired both barrels and it knocked me over, much to Leo's amusement. Another brother, Ed, was a crack shot. I was with him in a field when five prairie chickens got up and he shot them all with a Remington pump gun. Years later Ed came to watch us hunt. As a North Dakota resident, he didn't have a license. He complimented me on my shooting, which was high praise from a master. He did say I shot too quickly.

The Parker Brothers gun came to me, and I think it was the first gun my son Tim shot. Unfortunately it was stolen from our Moorhead house before Tim could make much use of it. Later I bought a Parker Brothers replica but it's not quite the same.

In the late 1920s the older brothers all left to do something better. My father didn't discourage them, since farming no longer seemed to offer much livelihood. Joe and Leo had worked for years with practically no wages. My mother was so fair. She gave Joe her inheritance of $1000 so he could get married, and she sent Leo the entire $700 of profit she had made on turkeys. The turkey business was mostly a disaster. You looked cross at one of them and half the flock fell over dead. Later she got into breeding stock and made some money selling prize toms.

With the older brothers gone we had to depend on hired men. These were usually sturdy fellows who were good workers and wanted a paycheck without responsibility. They educated me on the basics of life and I found out later they were pretty accurate.

We became a close-knit family, we three children, my mother and old father. With the added expense of the hired men things got progressively worse. My father's old Buick gave out. He was so proud of that car. We used to drive to church in Ada with the Barrys. Bill Barry was my father's age and had come from New York shortly after he did. The Barrys had a Dodge touring car just like the Buick. Both cars would go 66 miles an hour top speed, no more, no less. After church we would race home for fifteen miles. The two old men would drive abreast on the loose gravel road with the children yelling and laughing and neither one of them could get ahead. Our mothers were distraught but we never did have an accident.

I think the Dodge was William Barry's only indulgence. He didn't work and he always had money because he didn't spend any of it. He bought grain at a low price and sold it when the price was high. He never borrowed. Ed Murphy borrowed money to make money but most of the time things didn't work out. His excellent credit rating did him in.

It was 1933, the heart of the Depression that had begun with the Crash of 1929. We had a one-room red brick bank in Felton and I remember the day it closed and the ominous look of the drawn shade over the window. I asked my father if he had lost all his money when the bank closed. He answered, "No, I didn't have any to lose." And then he smiled and added, "I was way ahead of that game."

One day I came upon my father weeping. I said, "What's the matter, Papa?" He answered, "I'm a failure as a father. I brought you three into the world thinking you and your mother would be well cared for, and now it's all gone. I owe every friend I have and I even had to take the cash out of my life insurance. Everything I've tried has turned to ashes." I tried to cheer him up but he was inconsolable.

Shortly after that, at age seventy-seven, he had a stroke. We had lost a horse with sleeping sickness that year and I thought that was what it was. But it was a stroke and though he lived another year he was never the same. Thank God his worries were over. He believed the granaries were full and there was cash in the bank. This heaped everything on my poor mother, including his care. I began to realize I would have to become head of the family at age sixteen.

There was no rain that year. The ground became so dry that the least little bit of wind started a dust storm. The ditches were filled with pulverized soil. The cattle stood with their ribs showing and their haunches to the wind. They had no hay to eat and little water because the wells were running low.

One day I was walking behind four horses and a drag while the dust blew. I could hardly see and every muscle in my body ached. When I paused near the highway I saw a man go by smoking a cigar in a big black car. I had been reading Dos Passos and this prompted me to call the driver a bad name. I was leaning toward socialism and thinking there must be a better way. Years later I bought the longest, blackest Oldsmobile 98 to show there was no bitterness in my heart. I never did take up cigar smoking.

I never had a bicycle, but we had three Shetland ponies. These were generally ignoble little beasts who were tricky and not very friendly. If you wanted to ride, it was almost more trouble to catch them than to walk where you needed to go.

The horses were great animals—gentle, hard-working, long-suffering. We had a myriad of them. Greys, bays, sorrels, broncos, and blacks with all sorts of names: Dobbin and Daisy, Heck, Joe, Ole, Dan, Toots and on and on. We had two riding horses that were smaller bays, Pat and Fly. During the fall tilling I rode alone to the other farms two miles away. I would feed hay and oats to all the horses there, then I would curry and harness eight of them, lead them out four at a time and hitch them to a three-bottom plow. Russian thistles would plug up the plow and there couldn't have been a worse job devised by man. The ground was so pulverized that the mold boards wouldn't scour. At noon I would unhitch to feed and water the horses. I remember one noon I had hard boiled eggs and egg salad sandwiches for lunch. Afterward, the routine resumed until sundown when I settled the horses and rode the little mare home.

The fun in my life was going to school. Sometimes I had to skip and work on the farm, but my mother would not let that happen very often. We lived two miles north of the school, and we went back and forth in a rudimentary horse-drawn coach. It had four steel wheels, and when there was snow it was put on a sled. In the winter it was very cold and we were given heated bricks and horse blankets to keep us warm. No one minded the cold or privation because we all shared it. That's why it was great to grow up in such a time.

A new principal came to town. His name was Allen Erickson, and he was also a good teacher who introduced me to American literature. Despite the disparity in our ages, we became fast friends. He saw the coat I was wearing and bought me a new one, something I will never forget. Our friendship lasted for his lifetime.

There were so few boys that I was playing varsity basketball when I was in seventh grade though I was barely five feet tall. One year we played Moorhead in a tournament. They had been State Champs the previous year and were about to repeat the performance. We got the opening tip and a basket, so the score was 2 to 0 Felton. There was a center jump in those days, and we never got the ball after that. The score ended 72 to 2 Moorhead. My father said we were the "hind end" of America. He couldn't stand to lose.

The poor soul died on June 1, 1934, the day I was to graduate from high school. It was hot and still—hadn't rained and it seemed like the end of everything. There was a host of relatives and friends and it was a joyless celebration of death. A neighbor lady loaned her Buick Landau so we could ride from Felton to Ada for the funeral, then to Georgetown for the burial, and finally home. It was a long journey. My mother, who had been such a brick, went to pieces—but not for long.

The day after the funeral the creditors started coming. I couldn't believe the numbers. They were mostly decent, friendly people who had a high regard for my father. My mother had regained her composure. She greeted everyone in a cool and matter-of-fact way. She said we were going to negotiate a Federal Land Bank loan and to be patient and they would get a settlement.

—*Vincent Murphy*

MARSH GRASS

1992

Red Like Him

for R.P.W.

He was tutor to a lad
he never really knew—
only the shock of red
like sunrise on a slough.

Out for an autumn walk,
I hear the great geese cry
and hail a red-tailed hawk
spiraling in the sky.

Robert Penn (Red) Warren, who taught me poetry at Yale, once admonished me: "Go home, boy. Buy a farm. Sink your toes in that rich soil and grow some roots." At graduation I could hardly imagine returning to North Dakota. I was already a published poet, and I was drawn to the literary life of New Haven and New York. For a time I considered seeking the post of poet-in-residence at a prestigious academy. But Warren refused to recommend me. He said I'd already spent enough time among juvenile minds. Perhaps he realized my passion for academic life would prove as short-lived as the other love affair which kept me in Connecticut my first year out of Yale. Slowly I began to realize the old man was right. I needed to cultivate the sense of place which I so fervently admired in Yeats, Hardy, and Frost, but which I had not yet found in the land of my own birth.

Although the term "provincial" is often used as a slur, Warren was well aware that much of the best poetry *is* provincial, that it is written in and about Wessex or Ireland, New England or his own Kentucky. No doubt it amused him to introduce the red-haired "boy from Dakota" to his urbane guests, but he sensed all along where my life and my writing ought to go, though he might not have realized how circuitous a path I would have to take.

Warren was but one of the teachers who shaped my view of life. Among the others were my father, the only one still alive, and three master farmers I have helped to inter: Ed Benedict, Ivan Miller, and Hank Peterson. Surely no office is more sacral or more central to my sense of place than the burial of the dead.

For nearly twenty years my principal (and principle) business was death and taxes. My task was to help plan for the inevitable, then deliver the insurance proceeds and succor the survivors as best I could. Sometimes there was no policy to place: Hank Peterson was too old for insurance when I planned his estate. But he put the fear of God into younger farmers who had no wills or trusts, then sent me to their kitchen tables. They were willing to trust a red-headed stranger only because they trusted Hank. As financial advisor and confidant to these families, I have attended more funerals than I care to count. Like anyone else, I have also watched the passing generations of my own kin.

The Pallbearers

At the prairie cemetery
where the river meets a road
and Murphys come to bury
love in the loam we've sowed
my brother lets me carry
the light end of the load.

Despite our age difference of nearly fifty years, Hank Peterson was a friend and fellow hunter as well as a client. He was also known among his peers as an incorrigible practical joker. On one occasion, after he and Ed Benedict joined a party of deer hunters in northern Manitoba, Hank started kidding Ed about his shooting. Everyone knew that Ed was the best shot in camp, but Ed was so vain about his marksmanship that he took the bait and defended himself. "Tell you what," said Hank. "That's a nice new hunting cap you're wearing. If you can hit it at a hundred yards, I'll replace it." Ed angrily agreed, and Hank made a show of pacing off the range, to distract his audience so no one would notice that he was slipping a stick of dynamite under the cap. He got clear in a hurry once he set it on a stump. Ed's shot must have spooked half the deer in Manitoba. Hank counted the cap a small price to pay for the look on Ed's face after the explosion.

Hank didn't hesitate to play his tricks on me. Once I paid him a visit to propose he purchase a variable annuity. He wasn't in a buying mood when I arrived. There was a kerchief wrapped around his head and he lay groaning on the couch. "What happened to you, Hank?" "Went to the dentist," he slurred. Then he handed me a jar containing the two biggest molars I'd ever seen. Thinking of my own impacted wisdom teeth, never removed, I must have turned several shades paler. Errand forgotten, I stammered my condolences and fled. Later that afternoon Dad asked how I'd fared. He burst out laughing when I told him. "So he pulled the old sheep's tooth trick on you too."

For all his playfulness, Hank was dead serious about farming. He was far ahead of his time in perceiving agriculture as a business, and he also took a keen interest in agronomy. For many years he ran field trials of new seed varieties bred at North Dakota State University, and he funded scholarships for promising agronomists. During the 1960s Hank took to hectoring the Nobel Prize committee. He wanted the Peace Prize awarded to an obscure scientist who had devised the sterilization technique which eradicated the Mexican screwfly. Long forgotten now, this insect was a scourge of cattle and humans in Central and South America. Its exterminator, Hank reasoned, had done more to improve the lot of humanity than many another recipient. Sadly, the committee didn't agree.

Among Hank's many agricultural interests was the largest truck farm in the Red River Valley. With our four-month growing season, vegetables are an even dicier proposition than grain farming. Hank was trying to start a son-in-law in the business, but the younger man lacked the steadiness requisite for those who would wrest a living from the land. One chilly September evening the son-in-law burst in crying, "We've got 160 acres of unpicked tomatoes out there and it's going to freeze. What the hell should we do?" Hank replied, "I don't know about you, but I'm going to bed."

Implicit in Hank's remark was the uniquely American blend of resignation to the exigencies of a harsh land and confidence that any setback can be surmounted by a man of character. The frontier people observed by de Tocqueville and Crèvecoeur possessed that spirit in abundance, but now it seems depressingly scarce.

In a flush of agrarian enthusiasm, Thomas Jefferson once wrote, "No occupation is more delightful to me than the cultivation of the earth." On another occasion, probably after a blight or a hailstorm at Monticello, he observed, "In the lotteries of life . . . even farming is mere gambling."

Nowadays lotteries are sponsored by the states. Lining up for their tickets, city people get a taste of the expectation a farmer feels every time a crop goes in and the disappointment he suffers when the harvest fails to match his hopes. But a farmer in Arizona once won a fifteen-million-dollar lottery prize. Asked what he would do with the proceeds, he replied: "Keep farming till she's gone."

Sometimes I wonder whether Jefferson's Republic has lost something essential to its well-being in this age of Lotto. Would modern Americans take axes to crates of over-taxed tea, or construct, compromise by compromise, the fairest system of governance humanity has yet contrived? Heaven help them, if anything goes seriously wrong with the unseen systems of agriculture, transport, and finance that sustain their lives.

A Farmer's Prayer

Spirit of the wheat
brush every beard
turning green flaxen
with a wave of your wand.
The wind is your oven,
the hills your loaves.
Dry husks rustle,
flag leaves furl,
heads curl earthward
as kernels harden.
Your garden is golden,
your larder laden.
Feed a hungry world.

Farmers' prayers are rarely answered. Over time the strain of perpetual uncertainty takes a toll. People cope in different ways, and one way is drink. Hank was an alcoholic who sobered up about the time of my birth and founded the first chapter of A.A. in the Valley. A quarter century later, old age was wearing out his resolve. Lamed by an automobile accident, he was too tired to farm, too feeble to hunt. In despair, he turned again to his old friend, Jack Daniels. Drunk, Hank suffered a massive stroke which should have killed him. Thanks to the mercies of modern medicine, he survived to spend seven years paralyzed in a nursing home and unable even to speak. The last time I visited him I brought plat maps of Wilkin, Sergeant, and Ransom counties where I was buying land and starting my own farm venture. The nurses who had thought him utterly oblivious and unresponsive were amazed to see tears rolling down his cheeks.

I too took solace by the bottle when my fortunes turned bad. Losing those farms was the most gut-wrenching experience of my life. But I was able to convey them into the hands of friendly investors. Although the titles passed on, the Millers kept farming that land. My brother and I were able to save several tracts totaling seven hundred acres; and our downsized holdings, less burdened by debt, just about break even when the crops are poor.

I still drive through that country to hunt every fall, and I cannot look without some heartache on the fields I once owned. Theognis's laments for his lost patrimony mean more to me now than they did in my student days. But anyone who lives by the land knows that gain and loss are equally transitory. Severe winters have killed off the pheasant flocks we nurtured in milder times; but the heavy snowmelt has filled every dip with water and cattails, so our duck populations are hitting record numbers. One day, no doubt, drought will return; the sloughs will recede, and pheasants breed again in weedy fencerows.

Little Heart Butte

Grouse peck at its breast
and pheasants at its foot.
Buffalo berries west
and Russian olives east
girdle this shortgrass butte,
this table set for a feast.
I, the unbidden guest,
have little heart to shoot.

When I decided to keep hunting dogs, I didn't anticipate how their brief life spans would affect me. Soon after Dee died I left the hip-deep snow and bitter cold of another interminable winter to sail for a week in the Bahamas with two young friends and one old one. When I returned to North Dakota, I found Maud unaccountably pregnant. Misrecalling the time she was last in heat, I took her to our veterinarian, Dan Treat, an exceptionally kind man whom we call St. Francis of Fargo. Dan palpated her and assumed she was at an early stage of gestation. Labs tend to bear big litters, and I didn't want twelve puppies of indeterminate breed whelped in my basement. An abortion seemed the only option. But sonograms are not routine for dogs, so the vet and I were equally astonished when Maud's induced labor produced only two pups, which were nearly full term. One of them survived delivery and looked like a pure-bred Lab. After a day at the hospital, he was still alive, and Dan thought he might pull through. I hastily prepared a whelping box and brought the proud mother and her offspring home. The puppy didn't make it through his second night. In the morning Maud lay with her chin over his little corpse, trying to hide the evidence of her shame, as though she were to blame for the puppy's death.

For a whole year Maud clung and whined so piteously that I finally relented and bought another puppy from Gun Dog Kennels. Like all dogs born there, his AKC name begins with Elmwood, Carl's township in Sabin, Minnesota, where the elms are now only barkless skeletons along the Buffalo River. So the full name of Maud's new kennel-mate is Elmwood's Edwin Benedict. But given his taste for newspapers, boots, garden hoses, weed whackers, almost any object in the path of his teeth, he quickly acquired the nickname Shreddie.

Maud recovered her equanimity within hours of the puppy's arrival. Her pack had returned to its proper size; the empty kennel next to hers was occupied again. I soon resumed the drudgery of training, one dog howling jealously while I worked the other. Once more the orchard echoed with the blanks fired from a starter's pistol and the whistle blasts commanding "sit" or "come." Eddie hunted admirably at six months, retrieving dozens of doves and ducks. But the triumph of his first season took place in South Dakota, where Jimmy and I drove in search of pheasants, which we could no longer find near our own land.

Years ago a South Dakota friend introduced me to Lars Herseth, whose ranch borders Sand Lake Wildlife Refuge on the James River. Pheasants were still breeding successfully in the sheltered bottoms there, and most North Dakota hunters would have sold their souls for an invitation onto Herseth's land. At the edge of an overgrown feedlot, Eddie flushed his first rooster. My brother dropped it, but the trajectory of its last flight took it across the fence into the refuge. Jim hollered back, "Bring Maud," and I replied, "Throw the puppy over the fence." The summer of training had worked its magic: Eddie had marked where the bird fell, and he brought it back only slightly the worse for his teething. Jimmy climbed the fence and passed the puppy back to his proud owner.

Our father still cherishes the idea of hunting, but he rarely joins us in the field any more. At eighty, he can no longer lift a shotgun to shoulder height. Maud too is getting gray, and after a few retrieves her legs are so stiff I have to lift her into the truck before we can move on. By the time Eddie gets old and lame, I will probably lose the endurance for long drives at 5 A.M., for hard slogs over rough ploughing and decoy sets in early snow.

Elegy for Diktynna

Go if you must and swim
the dim waters of Acheron
    for Actaeon.

When my engraved grouse gun
passes to someone else's son
    I'll whistle "Come."

On first reading "A Dog Young and Old," the four-part poem whose title plays off "A Woman Young and Old," my poet friend Timothy Steele quipped that I had the advantage of knowing my dogs better than Yeats knew women. In reply I might have quoted another Irish poet, Patrick Kavanagh, who once wrote, "I know nothing of women, nothing of cities." Although I have written in some detail of my grandmother, Tessie, I've overlooked other remarkable women who ought to figure in this narrative.

Francis Miller, Ivan's mother, was a schoolteacher like Tessie, and their lives were remarkably parallel. Though I'm sure Francis took some pride in the fifty-thousand-acre agricultural empire amassed by her sons, she was aghast that they, her grandsons, and her great-grandsons all showed such indifference to formal education. Her perusal of their report cards was a terrifying experience for all three generations of Miller boys.

Like the women in many farm families, the Miller women seem less inclined to question their religion than the men, who tend to feel like Job whenever some new tribulation is visited upon the land. Faith affords them a strength and patience the men often envy. Two days after Ivan's death, I met with his sons, his brother, his bankers, and his widow, Jane. Her composure in the face of sudden and unexpected bereavement, her resolution as executrix and trustee, reminded me of Tessie facing down the creditors who so nearly claimed the last of her land but could not rob her of dignity.

Peggy Benedict in her eighties retains both dignity and a frail residual beauty. For fifty years she bore Ed's roguishness with aplomb and humor. Any farm wife of her generation regarded the cellar, with its racks of preserves and row of washtubs, as her personal domain, so she ignored Ed's complaints about her habit of leaving oddments on the darkened stairs. Once when he stumbled on her boots, he shouted, "Dammit, Peggy, if I'd fallen down these stairs and broken my neck, what would you have done?" She blithely replied, "I would have called Vince Murphy and told him to bring over the check."

Mother's Day

Mothers gathered from miles around
when the mare foaled on Mother's Day.
One lady's bonnet blew away
and cartwheeled over the trodden ground.
The mare ate it without remorse,
then groomed her newborn's glossy coat.
He whinnied as if to clear his throat:
"Excuse me. I am a little horse."

In the 1980s my horse-loving neighbor used to ride around the orchard or pass along the half-mile trail I maintained through the river woods. That trail is gone now, and the neighbor's house has stood empty for nearly a year, like many others along the Red. I saw two big floods in my youth, in 1967 and 1969, when the river reached its highest levels of the century. We are all aware, in the abstract, that our Valley is the clay-lined bed of a glacial lake; but none of us expected to see the lake try to reclaim its bed in our lifetimes.

Perhaps we should have recognized the peril implicit in our long run of snowy winters. We welcomed the record blizzard of January 1989, which broke the drought and restored our confidence that the Valley was not destined to relive the Dust Bowl. In the early 1990s old marks for annual snowfall fell twice. By 1994 I was moved to write the premonitory poem "Twice Cursed." But each spring the snowpack melted harmlessly. The Red would run high for a while, then another damp summer would scab my apples and rust my wheat.

In the fall of 1996, cold weather came early, and the ground was already hard frozen when the first big storm dumped more than a foot of snow in mid-November. Clay isn't very permeable at the best of times, and turns tough as slate when it freezes. The first precondition for a flood was hidden under the new snowpack.

Storm followed storm for the next four months. Sometimes the snow fell gently, festooning trees and burdening roofs. More often the wind blew, schools closed, and country towns hunkered under siege for days on end. A climactic blizzard swept sixteen inches of powder over the Valley on the thirteenth of March. Drifts piled twelve feet high. At forty inches on the level, the snowpack was the deepest ever measured.

Three days later the thaw began. It was slow at first, and people dared to hope we might somehow slip through without a disaster. For almost three weeks the weather stayed sunny and dry. Each afternoon the great drifts shrank a bit. Each night they refroze. Bare ground began to reappear where wind had scoured the snow. We were glad to see grass again, even if it was only last year's thatch; but no water was flowing into the river yet. It was all caught in the mini-glaciers along the shelterbelts, or dammed in frozen ditches.

Twice Cursed

Bristling with fallen trees
and choked with broken ice
the river threatens the house.
I'll wind up planting rice
if the spring rains don't cease.
What ancestral curse
prompts me to farm and worse,
convert my woes to verse?

April 5 dawned dim and dark as if a thunderstorm were coming. Even before the rain, I could hear water starting to run in the culvert across the river. Soon I was helping to sandbag around the well that served several households near the orchard. We built high enough to fend off a forecast crest equal to the hundred-year flood of 1967. At that level water would touch the back of my house and put my access road under, along with the orchard and miles of surrounding countryside.

All afternoon the rain fell and the river rose. No one had ever seen so much rain at this time of year. By dusk a hard north wind brought a sudden chill and sleet started beating against my windows. All over Fargo the lights flickered on and off. Just to the west an ice storm was taking down feeder lines. A hundred thousand cattle were freezing to death on North Dakota ranches.

April 6 brought full blizzard and a windchill of minus forty. Still frozen, the river crept up through its fringing woods like a glacier. Nothing could stop the flood now. When the storm finally passed, a deathly Arctic silence fell. On the night of April 8, I watched the Hale-Bopp comet glint balefully on the slurry of ice and water that was starting to edge across the road. At that moment, even though the forecast crest had not been raised, only postponed, I knew for certain we were facing an unprecedented natural disaster.

The next day my sump pumps failed and the basement started filling underneath me. Sheets of tinkling ice were spilling over my driveway. I had already parked my Bronco by the highway to town. Alone at the end of an oxbow, my house would soon be cut off. Up the road my neighbors were milling about in confusion, some fighting to save their houses, others taking flight with truckloads of belongings.

I had run the dogs out to Carl Altenbernd the previous day. He lived on slightly higher ground. If he flooded, we'd all wind up in Hudson Bay. Now I was a refugee too, taking shelter at my lawyer's house in downtown Fargo. As a property developer, I felt that was the safest place for me.

Jim was in his element: freezing water. The skills he learned on Canadian rivers proved invaluable during the flood. He saved our father's house single-handed, bearing supplies and extra pumps in his canoe. At one point a network news crew showed up in the area, and Jimmy paddled a cameraman out to an older, lower house that was just going under. Two hours later he enjoyed fifteen seconds of fame on national TV.

For a time, slowed by the cold wave, the Red receded slightly. I was able to send some new pumps out to my house. Then came the real thaw and with it, the real flood, which lasted another week. As the Red rose, a tributary river, the Sheyenne, burst its banks some miles southwest of Fargo and spread cross-country to attack the city from behind. A fifty-square-mile moving lake lapped across Interstate 29 into Fargo's new suburbs.

I drove past National Guard checkpoints as Fargo made its last stand, shifting forces from downtown levees to dike the unprotected flank. A huge fleet of earthmovers was tearing up the muddy fields between developments. Outside the wall, teams of volunteers were sandbagging whole neighborhoods. Back at my refuge, I wondered whether we would be safe even there, or whether we would be swept off in some unimaginable tide of refugees.

A lucky wind shift gave the dike builders time to finish their project. The next day, under warm spring sunshine, the Red paused for many hours; then slowly, slowly it began to recede. All that water was heading for Grand Forks, a smaller city with fewer resources and less vigorous leadership. The first crest, dammed by ice in the river channel, was overtaken by the second, and the merged floods burst over all defenses. Two days later the Red at Grand Forks peaked *five feet* above its record level there. At that point the river was running fifteen miles wide.

Next Year, Drought

I. The Frozen Flood

The blizzard hurtles southward
on a river swirling north,
and our oldest farmers say
the Red is not the strongest,
the longest or the broadest
but by God it's the coldest
meanest river in the world.

II. Symphony of a Thousand

I'm listening to Mahler
while students fight all night
to raise the city's dikes.
Decorate them for valor
whether we stand or not
when the crescendo strikes.

III. Going Under

Fargo's victims are weeping
for the people of Grand Forks.
Our steeple bells are pealing
and our believers kneeling,
but nothing stops the seeping,
not Christ or all his clerks.

IV. The Recovery

Thanks for drains and sumps,
for six-inch diesel pumps
and concrete septic tanks,
for neighbors closing ranks
to overcome the flood.
Thanks for our forebears' blood
which still runs in our veins
as rivers vein these plains.

All told I was a flood refugee for twenty days. At the crest, with ice floes and fallen trees storming downstream, no one could reach my place. For three days I had no idea what was happening there. Finally Jim ventured out. He found the new pumps still running. The river had invaded my basement and garage, but stopped four inches short of my main floor.

It took months to clean up, but scars on the psyche outlast the scars on the countryside. Next week I shall be forty-seven, which actuaries tell us was the male life expectancy in this country a century ago. Feeling my age, I've taken to spending more and more time away from the fierce prairie winters that delighted me as a boy.

Today I'm finishing *Ploughshare* aboard my catamaran, *Catullus,* which is berthed at a marina in Key West. Six blocks west of my dock is the Porter House with its Frost cottage where the old poet fled New England's winters to write about New England. Eight blocks south and "*very* near the cemetery" is the house where Richard Wilbur does the same.

Altogether elsewhere, in snow-covered barns, Rich Bell's pigs are converting corn to bacon and making my livelihood. Since I bought my first farm, everything I feared has come to pass, but most everything I hoped for has happened too.

Eidyllion

for Charlee

I am selling my farms
to build a butterfly barn
where multicolored swarms
will storm the glassy dome
to greet the midnight sun,
and that will be my home.

Notes

p. 18. *Barrows* are castrated pigs.

p. 50. My thanks to the Scots poets William Neill and Gerry Cambridge for their assistance with this poem. In modern English, "Passel o' Pups" runs as follows:

> Pretty babies, black and fine,
> with your eager sucks and tugs,
> will you be good hunting dogs
> worthy of your father's line?
> Will you make haste and take them down,
> frantic partridge, crouching grouse
> and the Devil's own pheasant flush?
> Will you smite the rabbit brown?
> Like the gray deer you must leap
> over that stony, weedy hill
> where the wind blows loud and shrill,
> sore and cold, where very deep
> drifts the snow. Drink your fill,
> glossy beasties. Suck and sleep.

Index of Poems

Index of Woodcuts